GEOGRAPHY

OF

HINDOOSTAN

GEOGRAPHY

OF

HINDOOSTAN

DESIGNED AS

AN APPENDIX

TO

THE USUAL WORKS ON GEOGRAPHY

GYAN

Published by

Gyan Publishing House
5, Ansari Road
Daryaganj, New Delhi-110002
Phone: 011-47034999, 9811692060
E-mail: books@gyanbooks.com

Distribution Network
gyanbooks.com
India, USA, Canada, UK, Australia

ISBN: 978-93-7133-546-1 (PB)

First Published, 1842

2nd Impression 2025

Printed at: Gyan Press, Delhi.

GEOGRAPHY OF HINDOOSTAN, DESIGNED AS AN APPENDIX TO THE USUAL WORKS ON GEOGRAPHY

GEOGRAPHY

OF

HINDOOSTAN,

DESIGNED AS

AN APPENDIX

TO

THE USUAL WORKS ON GEOGRAPHY

PUBLISHED FOR SCHOOLS

IN EUROPE.

Calcutta:

PRINTED AT THE SCHOOL-BOOK SOCIETY'S PRESS, AND SOLD AT THE DEPOSITORY, CIRCULAR ROAD.

1842.

ADVERTISEMENT.

The following Geography of Hindoostan was printed as an Appendix to Goldsmith's Grammar of Geography; but many persons preferring other elementary geographical works, for the convenience of such it has been thrown off separately.

Calcutta, April 19, 1834.

1st ed. 1000 *copies.*
2nd ed. 4000 *copies.*
3rd ed. 5000 *copies.*
4th ed. 5000 *copies.*

GEOGRAPHY

OF

HINDOOSTAN, &c.

Question.—Where is Hindoostan situated?

Answer.—In the south-eastern quarter of Asia.

Q.—Between what degrees of latitude and longitude does it lie?

A.—Between the 8th and 35th degrees of north latitude, and the 68th and 92nd degrees of east longitude.

Q.—What are the boundaries of Hindoostan?

A.—The Himalaya mountains on the north and north-east; the Burman empire on the east; the river Indus and the Indian Ocean on the north-west and west, and the Indian Ocean on the south.

Q.—What is the population of Hindoostan?

A.—About 134,000,000.

Q.—How may this population be divided?

A.—Considering the two castes of Joyn and Sheek as Hindoos, the proportion will be 12 parts Hindoos, three parts Mahommedans, and one part other tribes and sects.

Q.—What are the names of the principal rivers of Hindoostan?

A.—The Indus, Ganges, Jumna, Sutuleje, Jhylum, Gunduck, Krishna, Nurbudda, Mahanuddy, Tuptee, and Cavery: besides many others.

Q.—Into how many parts is Hindoostan divided?

A.—Into four Parts; 1st, NORTHERN HINDOOSTAN; 2nd, HINDOOSTAN PROPER; 3rd, The DECCAN; 4th, INDIA SOUTH OF THE KRISHNA.

Q.—What are the extreme length and breadth of Hindoostan?

A.—One thousand nine hundred British statute miles from north to south, in length, and one thousand five hundred from east to west, in breadth.

1st, NORTHERN HINDOOSTAN.

Question.—What are the boundaries of this division?

Answer.—The Sutuleje river on the west; the Teesta river on the east; the Himalaya mountains on the north; and Hindoostan Proper on the south.

Q.—What are the subdivisions of Northern Hindoostan?

A.—1st, The country between the Sutuleje and Jumna; 2nd, Gurwal or Serinagur; 3rd, The sources of the Ganges; 4th, the Kumaon; 5th, Pamkhandi; 6th, Bhutan; 7th, The dominions of Nepaul.

2nd, HINDOOSTAN PROPER.

Question.—What is the northern boundary of Hindoostan Proper?

Answer.—North Hindoostan.

Q.—What is its southern boundary?

A.—The Nurbudda river.

Q.—How many provinces are there in Hindoostan Proper, and what are their names?

A.—Thirteen, viz. Bengal, Behar, Allahabad, Oude, Agra, Delhi, Lahore, Cashmere, Ajmeer, Mooltan, Cutch, Gujerat, and Malwa.

PROVINCE OF BENGAL.

Question.—In what part of Hindoostan is this province situated; and between what degrees of latitude?

Answer.—Towards its eastern extremity, and between the twenty-first and twenty-seventh degrees of north latitude.

Q.—What are the boundaries of this province?

A.—The dominions of Nepaul, Sikkim, and Bhutan, on the north; the Bay of Bengal on the south; Assam and Burmah on the east; and Behar on the west.

Q.—What may be its estimated length and breadth?

A.—Three hundred and fifty miles, or about 307 koss long; and 300 miles, or about 263 koss broad.

Q.—What is a koss?

A.—An India measure, equal to about 1 and $\frac{7}{8}$ British statute mile.

Q.—What is the population of Bengal?

A.—About 25,306,000.

Q.—What are the languages usually spoken in Hindoostan?

A.—The original language was the Sunskrit, and the characters in which that is written, the Debnagur: but the Sunskrit has long been confined to the Brahmuns and Pundits, and various other dialects substituted for it, and which are written in characters almost as various. From the time of the Mahommedan incursions, about 800 years ago, many Arabic and Persian words have been mixed with the Bengalee and Hindee, the two dialects that retain the greatest resemblance to the Sunskrit; and from still further corruptions of these have sprung nearly forty different dialects.

BENGAL.

Question.—What are the names of the districts into which the province of Bengal is divided?

Answer.—Calcutta, Dacca, and Moorshedabad, at each of which courts of appeal are held.

Q.—What is the name of the chief city of Bengal?

A.—Calcutta, which is likewise the capital of all India, being the residence of the Supreme authorities both in Church and State.

Q.—On what river is Calcutta situated?

A.—On the east bank of the western branch of the Ganges, called by Europeans the Hooghly, but by the natives the Bhagirathi, and considered by them peculiarly holy.

Q.—What distance from the sea is Calcutta?

A.—One hundred miles. Its citadel, Fort William, stands in lat. 22° 23′ north, long. 88° 28′ east.

Q.—What Churches are there in Calcutta?

A.—The Cathedral of St. John, and three other Churches belonging to the Establishment; one belonging to the Presbyterian Church of Scotland; three Roman Catholic Churches; one Greek; one Armenian, and several Chapels, belonging to different denominations of Dissenters. There is also one small Jewish Synagogue.

Q.—What number of Jews are there in Calcutta?

A.—The number of Jews in Calcutta does not exceed 250, and of these the larger proportion are very poor.

Q.—What Societies are there in Calcutta for promoting Religion and the Sciences?

A.—The Bible Society; the Society for propagating the Gospel in Foreign Parts; the Church Missionary Society; the Baptist Missionary Society; the Prayer Book and Homily Society; the London Missionary Society; the School-Book Society; three Societies for Native Female Education; the Asiatic Society; the Medical Society; and the Horticultural and Agricultural Societies, with many others.

Q.—What public buildings are there?

A.—The Government House; Her Majesty's Supreme Court-House; the Town Hall; the General Hospital; the Jail; the Writers' Buildings; the Hindoo College; the Mahommedan College; the European Female Orphan Asylum, and the Free School.

Q.—What other cities of note are there in Bengal?

A.—Dacca, the ancient capital of Bengal, and Moorshedabad.

Q.—What are the principal rivers in Bengal?

A.—The Ganges, the Brumhopootro, Pudma, Dummooda, Kooshee, Roopnarain, and several smaller rivers.

Q.—What are the productions of Bengal?

A.—Rice, sugar, salt, saltpetre, opium, indigo, tobacco, cotton, silk, and cloth, of a variety of descriptions; as also pulse and grain of various kinds; potatoes, cocoanuts, mustard seed, linseed, and many of the European vegetables.

Q.—Are there any mountains in Bengal?

A.—None; there are some rising grounds that are called hills, although more deserving the designation of inequalities; but the country, generally, is of a flat champaign description.

Q.—What are the prevailing winds in Bengal?

A.—In the south of Bengal the prevailing winds are north and south; in Behar, east and west; and the same in Assam; and these winds prevail according to the seasons of the year; which is divided into cold, hot, and rainy.

Q.—When do the periodical rains begin in Bengal?

A.—Early in June, and break up in September; they are succeeded by the cold weather, towards the middle of November, which lasts until about the end of February, when the hot season commences.

Q.—At what places are held the principal Courts of Justice?

A.—Calcutta, Dacca, and Moorshedabad.

Q.—Into how many zillahs are each of these circuits divided?

A.—Calcutta is divided into nine; Dacca and Moorshedabad into seven each.

Q.—What are the names of the zillahs (or districts) into which the circuit of Calcutta is divided?

A.—1st, Zillah Huwallee Shuhur; 2nd, the Twenty-four Pergunnahs; 3rd, Jessore; 4th, Hooghly; 5th, Nuddea; 6th, Burdwan; 7th, Jungle Mehals; 8th, Midnapoor; and 9th, Cuttack. Midnapoor and Cuttack were formerly attached to Orissa.

OF THE ZILLAH HUWALLEE SHUHUR.

Question.—In what province is this zillah?

Answer.—In the province of Bengal.

Q.—In what district?

A.—The district of Calcutta.

Q.—What is the number of Police Offices in it?

A.—Five, situated in the suburbs of Calcutta; viz. Chitpoor, Manicktola, Tajhat, Nuhazaree, and Sulkea.

Q.—What is the population of this zillah?

A.—The statements have been numerous and various, and there does not yet appear to be any very correct estimate of the zillah; but it may probably be fairly stated at between 600,000 and 700,000 souls in the city and its suburbs; and between 200,000 and 300,000 beyond the suburbs.

Q.—What may be the proportions of the several religious persuasions in Calcutta?

A.—In the year 1822, the following returns of the population of Calcutta, *exclusive* of its suburbs, were made:—Christians, 13,138; Mahommedans, 48,162; Hindoos, 118,203; Chinese, 414; making a total of 179,917.

Q.—What is the principal court of justice in this zillah?

A.—As regards Calcutta, that is, the territory within the Maharatta ditch, ceded to the British Crown, and all the British inhabitants of Bengal, the Supreme Court; the business of which is conducted by a Chief

Justice, and two Puisné Judges, who receive their appointments from the King of Great Britain.

Besides this, Calcutta is the seat of a Court of Appeal and Circuit, from all the courts in the several zillahs already stated, as comprising the district of Calcutta.

Q.—What is the extent of trade carried on in Calcutta?

A.—In 1814, the imports, from beyond seas, are stated at 18,100,000 Sa. Rs. The exports, at Sa. Rs. 47,600,000. The inland imports, and exports amounted together to Sa. Rs. 10,400,000, making a grand total of Sa. Rs. 76,100,000.

THE TWENTY-FOUR PERGUNNAHS.

Question.—In what province is this zillah situated?

Answer.—In the province of Bengal.

Q.—In what district is it?

A.—The district of Calcutta.

Q.—What are the boundaries?

A.—This zillah is bounded on the north by zillah Nuddea; on the south by the Bay of Bengal and the Sunderbunds; on the west by the river Hooghly; and on the east by the zillah Jessore.

Q.—What rivers are there in the zillah?

A.—The Hooghly, the Raymungul, and a number of smaller ones, that pass through the Sunderbunds to the sea; and also a small river, or as it is designated by the natives, a Nulla, named Catigunga, or Tolly's Nulla, which passes near Calcutta, and affords the means of communication, by water, between that city and Dacca.

Q.—What is the extent of this zillah?

A.—About 180 miles from north to south, and between 100 and 110 from east to west.

Q.—What is its population?

A.—About 1,200,000 souls, of which 12 out of 16 parts are Hindoos; the remaining 4 parts Mahommedans; there are many Europeans not included in the above statement. The Judge and the Collector reside at Allypoor, where the court is held.

Q.—What are the chief articles of trade in this zillah?

A.—Rice is cultivated to a large extent, as is also indigo; fruits, likewise, of various kinds, are grown in considerable abundance; and being near the sea, salt forms no small part of its produce.

Q.— What are the principal trading towns?
A.—Borahonugur, Nowabgunge, Bishnapoor, and Cologachee.

OF ZILLAH JESSORE.

Question.—In what province is this zillah?
Answer.—In the province of Bengal.
Q.—In what district is it?
A.—The district of Calcutta.
Q.—What are its boundaries?
A.—On the north, the great Ganges; on the south, the sea; on the west, the zillah Nuddea and the river Hooghly; and on the east, Dacca-Jelalpoor and Backergunge.
Q.—Between what degrees of latitude does Jessore lie?
A.—Between the 22nd and 23rd degrees of north latitude.
Q.—What are its dimensions?
A.—In the year 1814, it was estimated to contain about 5,000 square miles. In the southern part of this district are the Sunderbunds, a tract of low swampy country, intersected with rivers and inlets of the sea.
Q.—Are the Sunderbunds much cultivated?
A.—No; the land is fertile, but the unhealthiness of the climate, and the expense of embankments, to keep out the surrounding water, have heretofore proved serious obstacles to the advance of cultivation; but these are gradually giving way to the enterprize of European settlers.
Q.—What are the principal rivers of this zillah?
A.—The Bhoirob, Chittra, Nobo-gunga, Coomar, and the Modhoomotee.
Q.—Which are the chief towns of this zillah?
A.—Moorley, Bhoosna, Mahmoodpoor, Noldanga, Mirjanogur, Gopalgunge, and Coolnea.
Q.—Which of these is the capital of the zillah?
A.—Moorley, also called Jessore, situated in lat. 23° 7′ N., long. 89° 15′ E. and about 62 miles N. E. from Calcutta; the Government functionaries reside here.
Q.—What is the population of this zillah?
A.—The last census was taken in the year 1801, and the return made was 1,200,000, in the proportion of nine Mahommedans to seven Hindoos.

Q.—What are the most valuable articles of produce in this zillah?

A.—Rice, salt, indigo, (of an excellent quality) tobacco, ganja, mulberry, pawn, betelnut, and long-pepper. Cocoanuts are also grown in great quantities; matting and cloth are likewise articles of trade; and the herds of cattle are numerous and large.

HOOGHLY.

Question.—In what province is this zillah?

Answer.—In the province of Bengal.

Q.—In what district is it?

A.—The district of Calcutta.

Q.—What are its boundaries?

A.—It is bounded on the north by Burdwan and the river Hooghly, which separates it from the zillah of Kishnagur; on the east by the river Hooghly, which separates it from the district of Jessore; on the west and south by Midnapoor.

Q.—Between what degrees of latitude does this zillah lie?

A.—Between the 22nd and 23rd degrees of north latitude.

Q.—What is the general appearance and soil of this zillah?

A.—A low, flat, but fertile, country.

Q.—What are the principal towns in this zillah?

A.—The town of Hooghly is the capital of this zillah; it is situated in 22° 54′ N. lat. and 88° 28′ E. long.; and during the Mogul government, was a town of considerable note.

Q.—Were there, at any time, any other factories here?

A.—Yes; the French, Dutch, Portuguese, and Danes had each a factory here; and were subsequently permitted to build towns along the west bank of the river—the French at Chandernagore, the Danes at Serampore, the Dutch at Chinsurah, and the Portuguese at Bandel.

Q.—Is there any thing worthy of notice at Bandel?

A.—Yes, the Portuguese built here a splendid Roman Catholic Church, and established a Monastery and a

Convent. The Church is still in good repair, although bearing the date of 1592; a considerable part of the Monastery still remains, and is used for various purposes by the Vicar and his assisting Priests: but the Convent has fallen to decay; the garden, with its wall, being all that is standing.

Q.—What was the ancient name of Hooghly?

A.—The ancient name of Hooghly was Golin. It was captured by the Mogul army in the year 1632.

Q.—Is Hooghly still a town of any note?

A.—The town of Hooghly is now, comparatively speaking, an inconsiderable place, although the courts of the Judge and Magistrate and Circuit Judge are held in it.

Q.—What other towns are there in this district?

A.—Chunderkona, Oolubarea, Kheerpoy, and some others of inferior note.

Q.—What is the population of this zillah?

A.—In the year 1801, it was estimated at 1,000,000, but later statements make it 1,200,000, in the proportion of three Hindoos to one Mahommedan.

Q.—What are the chief productions of this zillah?

A.—Rice, sugar, indigo, and salt. Amongst the natives, the salt of this zillah is looked upon as possessing peculiar sanctity, as being extracted from the mud of the most sacred branch of the Ganges.

Q.—Are there any seminaries for education in this zillah?

A.—The only seminaries supported by the Government, now in this zillah, are a few schools, in which the children are taught to write, read, and cypher, in Bengalee: the commencement of a college has now, however, been made.

Q.—Were there, at any previous period, any other seminaries?

A.—In the year 1801, there were thirty students, instructed in the Persian and Arabic, at Seelapoor, at an institution maintained by the produce of lands, under a grant confirmed by Governor Hastings. At the same place, the principles of Hindoo law were taught by Pundits, in about 150 private schools, containing from five to twenty scholars in each: but these schools have fallen into decay.

ZILLAH NUDDEA.

Question.—In what province is this zillah ?
Answer.—In the province of Bengal.
Q.—In what district is it ?
A.—The district of Calcutta.
Q.—What are its boundaries ?
A.—It is bounded on the north by the Zillah of Rajshahee ; on the south by the Twenty-four Pergunnahs and Jessore ; on the east by Jessore ; and on the west by the river Hooghly, which separates it from Burdwan.
Q.—Between what degrees of latitude does it lie ?
A.—Between the twenty-second and twenty-fourth degrees of north latitude.
Q.—Of what nature is the soil of this zillah ?
A.—The soil is rich in all the dearer productions of India, although so light as only to bear cultivation for three successive years ; after which it requires to lie fallow for the same period.
Q.—What number of bigahs does this zillah contain ?
A.—It contains about 4,024,000 bigahs, out of which about 900,000 bigahs are rent-free. The proportion of cultivated to the uncultivated land was estimated, in the year 1802, as follows :

Jungle, ...	bigahs,	105,000
Ground lying fallow, ...		1,959,500
For cultivation, ...		1,959,500

Q.—What is the population of this zillah ?
A.—The latest statements made the population to amount to about 1,400,000, of which three parts are Hindoos and one part Mahommedans.
Q.—What are the chief articles of produce in this zillah ?
A.—Mulberry trees, hemp, flax, tobacco, sugar, and oush, (a plant from which a red dye is procured ;) also pulse, indigo, mangoes and various other fruits and vegetables.
Q.—Is much rice grown here ?
A.—No ; owing to the soil being sandy, it is less favourable to rice, a smaller proportion of which is, consequently, cultivated, than in other zillahs.

Q.—Are many cattle raised in this zillah?

A.—Yes; the soil being dry and sandy, cattle are abundant.

Q.—What are the principal rivers in this zillah?

A.—The Hooghly, Jellinghy, and Issamutty, all which afford an easy and speedy means of carriage for goods of every description.

Q.—What are the chief towns in this zillah?

A.—Nobodweep, Santipoor, Oola, and Koomarhatta: there are, besides, several smaller ones, viz. Sheebneebas, Krisnogur, Sooksagur, Ranaghat, Chagda, and Nuddea.

Q.—Is there any thing deserving of notice regarding the town of Nuddea?

A.—Yes, it was the capital of a Hindoo principality, previous to the Mahommedan conquest.

Q.—When did it become subject to the Mahommedan sway?

A.—It was taken, and entirely destroyed, by Mahommed Bukhtyar, the first Moosulman invader of Bengal, in the year of our Lord 1203, during the time that Cuttub ud Deen occupied the Delhi throne.

Q.—What is the situation of Nuddea?

A.—It stands at the junction of the Jellinghy and Cossimbazar branches of the Ganges, in lat. 23° 25′ N., long. 88° 24′ E. about 60 miles north of Calcutta.

Q.—Is there any thing else for which Nuddea has been remarkable?

A.—Yes, it was formerly the seat of Brahminical learning, being the site of a Hindoo College. In latter years, this has fallen into complete decay, and now there do not appear to be any seminaries within the district, in which either Hindoo or Mahommedan law is taught.

Q.—Where does the Judge hold his court?

A.—At Krisnogur.

Q.—What part of the zillah is Plassey situated in?

A.—In lat. 23° 45′ N. long. 88° 15′ E. about 30 miles south of Moorshedabad.

Q.—Is there any thing remarkable in the history of Plassey?

A.—Yes, it is famous for the adjoining plain on which was fought, on the 23rd June, 1757, the famous battle between the British troops, under Colonel Clive, consisting of 900 Europeans, 100 topasses, and 2,000 sepoys, with 8 six-pounders and howitzers; and the young Nuwab

Seraje ud Dowlah's army, consisting of 50,000 men, 50 pieces of cannon, and about 40 French fugitives from Chandernagore.

Q—What was the result of this battle?

A.—This battle, in which the British gained a complete victory, decided the fate of Bengal, and ultimately that of India.

BURDWAN

Question.—In what province is this zillah situated?

Answer.—In the province of Bengal.

Q.—In what district is it?

A.—The district of Calcutta.

Q.—What are its boundaries?

A.—It is bounded on the north by Beerbhoom and Rajshahee, from which it is divided by the river Adji; on the south by Midnapoor and Hooghly; on the east by the river Hooghly; and on the west by the Jungle Mehals.

Q.—Between what degrees of latitude does it lie?

A.—Between the 22nd and 24th degrees of north latitude.

Q.—What are its dimensions?

A.—In the year 1814, it was computed at 2,400 square miles.

Q.—What may be its population?

A.—In 1814, it was stated to be 1,444,487 persons; subsequent accounts make it not less than 1,450,000, of which thirteen parts are Hindoos, and three parts Mahommedans.

Q.—What is the nature of the soil?

A.—Remarkably fertile; perhaps more so than any other part of Bengal, or even of India. The inhabitants bestow much pains in its cultivation, for which they are repaid by an abundant return of valuable productions.

Q.—What are the chief articles of produce?

A.—Sugar, indigo, pawn, cotton, tobacco, grain, and mulberry trees.

Q.—What are the names of the principal towns?

A.—Burdwan, Kanchonnogur, Cutwa, Umbika, Mauncour, Mungulkoth, Culna, and some others.

Q.—Which of these is the capital of this zillah.

A.—Burdwan is the capital of the zillah, and in which is the palace of the Rajah of Burdwan, and adjoining it the residences of the Circuit Judge, the Judge, the Magistrate, the Collector, and other public functionaries of the East India Company; as also the cantonment of a Provincial Battalion.

Q.—What are the names of the principal rivers in this zillah?

A.—The Hooghly, the Adji, and the Damooda.

Q.—What are the latitude and longitude of the town of Burdwan?

A.—Its latitude is 23° 15′ N., its longitude 87° 57′ E. Its distance from Calcutta 60 miles N. N. W.

Q.—What is its population?

A.—It is stated to be 53,927 persons.

JUNGLE MEHALS.

Question.—In what province is this zillah?

Answer.—In the province of Bengal.

Q.—In what district is it?

A.—The district of Calcutta.

Q.—What are the boundaries of this zillah?

A.—Beerbhoom on the north; Midnapoor on the south; Burdwan on the east; and Ramghur and Chota Nagpoor on the west.

Q.—How long is it since this zillah was formed?

A.—About twenty years ago.

Q.—What zillahs was it formed from?

A.—It consists of dissections from the contiguous zillahs of Beerbhoom, Midnapoor, and Burdwan.

Q.—What is the name of the capital of this zillah?

A.—Bancoora, where the public functionaries reside, and the different courts are held.

Q.—What distance is Bancoora from Calcutta?

A.—About 98 miles; and it is beautifully situated on a rising ground, the soil of which is a coarse gravel.

Q.—What is the population of this zillah?

A.—The population is not as yet correctly ascertained.

Q.—Is there any thing worthy of notice in this zillah?

A.—The territories of the Rajah of Vishnoopoor are incorporated with this zillah; the Rajah's lineal succession held this ancient kingdom in possession for 1,100 years free of tribute.

Q.—What rivers are there here?

A.—The Damooda, Adji, Shilai, Dalkeesor, and Cosai.

Q.—Is there any thing else deserving notice here?

A.—Yes; a new road, making by the Government, between Calcutta and Benares, passes through this zillah.

MIDNAPOOR.

Question.—In what province is this zillah?

Answer.—In the province of Bengal.

Q.—In what district is it?

A.—The district of Calcutta.

Q.—What are the boundaries of this zillah?

A.—Jungle Mehals and Burdwan on the north; Balasore and the Bay of Bengal on the south; the Hooghly on the east; and Ramghur on the west.

Q.—What is the population of this zillah?

A.—According to the last statements 1,500,000. Two famines, one in the year 1770, the other in the year 1799, each of which carried off nearly half the inhabitants, have lessened the population of this zillah very much; but it is now gradually increasing.

Q.—What are the productions of this zillah?

A.—Sugar, indigo, betelnut, and cloth of a superior quality.

Q.—What is the name ofthe capital of this zillah?

A.—Midnapoor, situated in lat. 22° 25′ N. and long. 87° 25′ E. and 70 miles W. by S. of Calcutta.

Q.—What other principal towns are there?

A.—Jellasore, Piplee, Nagurgur, Tumlook, and Kedgeree.

Q.—Which was the first town at which Europeans traded in this zillah?

A.—Piplee, and then Balasore.

CUTTACK.

Question.—In what province is this zillah?

Answer.—In the province of Bengal.

Q.—In what district is it?

A.—The district of Calcutta.

Q.—What are the boundaries of this zillah?

A.—Midnapoor and Moyurgunge on the north; the district of Ganjam on the south; the Orissa territories on the west; and the Bay of Bengal on the east.

Q.—What may be the size of this zillah?

A.—It contains about 9,000 square miles.

Q.—How is it divided?

A.—Into two districts; north and south.

Q.—What is the name of the north district?

A.—Balasore.

Q.—What is the name of the south district?

A.—Juggurnauth.

Q.—What is the number of the population?

A.—About 1,200,000, of whom nearly the whole are Hindoos.

Q.—Are there any rivers in this zillah?

A.—Few of the rivers of this zillah have any current throughout the year; they are much swollen during the rains, but most of them dry up when they abate.

Q.—What are the names of the principal ones?

A.—The Mahanuddy, Brahminy, Byturini, Solandy, Kausbans, Buranalang, and the Subanreeka.

Q.—What are the names of the principal towns of this zillah?

A.—Balasore, Bhodrok, Cuttack, and Juggurnauth.

Q.—Is this a mountainous zillah?

A.—No, near the sea it is low and swampy; about 20 miles inland it rises a little, and becomes more fertile; and about 40 miles from the sea, it swells into hills, the highest of which, however, does not exceed 2,000 feet above the level of the sea; and their general elevation is from 300 to 1,200 feet.

Q.—Are these hills barren, or wooded?

A.—They are covered with trees; some resinous, and others valuable for the purposes of cabinet-work and dyeing; and they contain a great variety of valuable and curious minerals.

Q.—What are the productions of this zillah?

A.—Salt may be considered as its chief article of commerce; a large quantity of a superior quality of which is manufactured, yielding a revenue of above eighteen lacs of rupees annually. Rice, grain, pulse, aromatic roots, spices, dyeing drugs, and sugar-cane grow here; and in the hilly tracts, Indian corn, and wheat are also produced.

Q.—What is the name of the capital of this zillah?

A.—Cuttack; a town situated in lat. 20° 27′ N., long. 86° 5′ E.; 251 miles travelling distance from Calcutta.

Q.—What number of inhabitants are there in this town?

A.—About 40,000.

Q.—Is Cuttack an ancient town?

A.—Yes; it is said to have been a capital so early as the tenth century.

DISTRICT AND ZILLAH OF DACCA.

Question.—In what province is this district situated?

Answer.—In the province of Bengal.

Q.—How many zillahs are there included in this district, and what are their names?

A.—Seven, viz. 1st, the town of Dacca, with the country in its immediate vicinity; 2nd, Dacca-Jelalpoor; 3rd, Mymunsing; 4th, Sylhet; 5th, Backergunge; 6th, Tippera; and 7th, Chittagong.

Q.—What are the boundaries of the zillah of Dacca?

A.—Mymunsing on the north; Backergunge on the south; the Bramhapootra river, which separates it from Tippera, on the east; and the zillah of Dacca-Jelalpoor on the west.

Q.—Between what degrees of latitude does it lie?

A.—Between the 23rd and 24th degrees of north latitude.

Q.—What is the name of the capital of this zillah?

A.—Dacca; a large town, situated in lat. 23° 42′ N. long. 90° 17′ E.; by road from Calcutta about 180 miles; by water nearly 400 miles. The city became in the year 1608, the metropolis of Bengal, which, with some short intermission, it continued during the remainder of the Mogul sway. Towards the end of the seventeenth century it was the residence of Azim Ashaun, the grandson of Aurengzebe, who commenced and nearly finished a magnificent palace now in ruins.

Q.—What is the number of the inhabitants of Dacca?

A.—About 120,000, chiefly Mahommedans.

Q.—Upon what river does the city of Dacca stand?

A.—Upon the Booree Gunga, or old Ganges, which is here about a mile in width.

Q.—What are the productions of this zillah?

A.—Salt, grain, and tobacco.

Q.—What are the names of the chief rivers?

A.—The Bramhapootra, Booree Gunga, Dulsuraee, and the Luckhee.

Q.—What towns besides Dacca are there in it?

A.—Narayungunge and Soonargaon.

DACCA-JELALPOOR.

Question.—In what province is this zillah?

Answer.—In the province of Bengal.

Q.—In what district is it?

A.—In the district of Dacca.

Q.—What are the boundaries of this zillah?

A.—Zillah Rajeshahee and Mymunsing on the north; Backergunge on the south; the zillah of Dacca on the east; and Jessore on the west.

Q.—What may be the extent of this zillah?

A.—It contains about 6,000 square miles.

Q.—What number of inhabitants does it contain?

A.—About 800,000; seven parts of whom are Hindoos, and nine parts Mahommedans.

Q.—What are the productions of this zillah?

A.—Indigo, betel-nut, and a kind of cotton, named Bhaga; also khassa, dimity, and other cloths.

Q.—What are the names of the principal rivers of this zillah?

A.—The Pudmah; besides several others of less note.

Q.—What is the name of the chief town of this zillah?

A.—Fureedpoor, where the court is held, and the Judge and Collector reside.

Q.—What distance is Fureedpoor from the city of Dacca?

A.—About forty miles.

MYMUNSING.

Question.—In what province is this zillah?

Answer.—In the province of Bengal.

Q.—In what district is it?

A.—In the district of Dacca.

Q.—What are the boundaries of this zillah?

A.—The Garrow mountains, and the district of Rungpoor, on the north; Dacca-Jelalpoor on the south; Sylhet and Tippera on the east; and Rajeshahee and Dinajpoor on the west.

Q.—Between what degrees of latitude is it situated?

A.—Between the twenty-fourth and twenty-fifth degrees of north latitude.

Q.—Is this an ancient zillah?

A.—No, it has been established only about thirty-eight years.

Q.—What number of inhabitants does it contain?

A.—About 1,360,000, of whom, in the year 1821, the Mahommedans were in the proportion of six to ten Hindoos.

Q.—What is the nature of the soil?

A.—The soil is very fertile, owing to the annual inundations.

Q.—What are the productions?

A.—Coarse rice and mustard, which it yields in great abundance.

Q.—What are the principal towns?

A.—Nusirabad, where the Judge and Collector reside, and the court is held; and also Bygonbarry, Serajegunge, Bokanagur, and Caugmary; Serajegunge is a place of very considerable trade.

Q.—What are the chief rivers of this zillah?

A.—The great river Bramhapootra, and the innumerable streams flowing into it, intersect the whole zillah.

SYLHET.

Question.—In what province is this zillah?

Answer.—In the province of Bengal.

Q.—In what district is it?

A.—In the district of Dacca.

Q.—What are its boundaries?

A.—On the north it is separated from Assam by an extensive range of mountains, inhabited by the Kookees, Kacharees, Garrows, and other barbarous tribes: on the east two small hill states separate it from the Burmese empire; on the south is Tippera; and on the west Mymunsing.

Q.—What number of inhabitants are there in this zillah?

A.—According to the best estimates that have as yet been obtained, about 500,000.

Q.—What is the general aspect of the country?

A.—The eastern and southern portions are hilly; but

the northern, central, and western parts are flat, and inundated during the rains.

Q.—Have the British any possessions in Hindoostan, to the eastward of this?

A.—No, this is the most easterly of the British possessions in Hindoostan, and is within 350 miles of the province of Yunan, in China.

Q.—What are the productions of this zillah?

A.—Rice is produced in large quantities in the low lands; sugar and cotton are raised on the rising grounds adjoining the low lands, but in quantities scarcely sufficient for the supply of the inhabitants. Oranges and limes are the chief articles of exportation.

Q.—Does Sylhet carry on any trade with the neighbouring tribes?

A.—Yes, it deals in chunam, of which there is a great quantity here; also in wax and ivory with the mountaineers on the eastern frontiers.

Q.—Are there any elephants in this zillah?

A.—Yes, they are very numerous; and a great number are annually caught for the service of the Government; but they are considered inferior, both in size and quality, to those of a more southern latitude, and nearer the sea.

Q.—Is there any coal in this zillah?

A.—Yes, coal was discovered in large quantities in the Laour mountains, in the year 1814, but the quality was not very good, and although it has since improved, no extensive mines are worked.

Q.—What is the name of the capital of this zillah?

A.—Sylhet, a town situated in lat. 24° 55′ N., long. 91° 40′ E.; travelling distance from Calcutta about 325 miles.

BACKERGUNGE.

Question.—In what province is this zillah?

Answer.—In the province of Bengal.

Q.—In what district is it?

A.—In the district of Dacca.

Q.—What are its boundaries?

A.—Dacca-Jelalpoor on the north; the Bay of Bengal on the south; the river Megna on the east; and Jessore on the west.

Q.—Is this an ancient zillah?

A.—No, it was formed about thirty years ago from the southern quarter of the too extensive zillah of Dacca-Jelalpoor, and includes the island of South Shahabazpoor, which produces much salt.

Q.—What number of inhabitants does it contain?

A.—About 1,200,000; ten parts of whom are Hindoos, and six parts Mahommedans.

Q.—What is the nature of the soil?

A.—Notwithstanding the lowness of the land, the soil is very productive, being fertilized by the deposits of the Ganges during the annual inundation, occasioned by the periodical rains.

Q.—What are the productions of this zillah?

A.—Rice is its staple commodity, of which it produces two abundant crops during the year; a large proportion of which is exported to supply the Calcutta market.

Q.—What is the name of the capital of this zillah?

A.—Backergunge was until the year 1801, when the courts of justice and revenue, and the residence of the Judge, Collector, and other public officers were removed to Burrisal, which has in consequence become the capital.

Q.—Where is Burrisal situated?

A.—Burrisal is in lat. 22° 46′ N., long. 96° 17′ E.; about 72 miles south from Dacca, and stands on the point of an oblong island, formed by the broad branches of the great Ganges, which is here immensely wide, and affords great facilities for inland navigation.

Q.—What is the situation of Backergunge?

A.—It stands in lat. 22° 42′ N., long. 89° 20′ E., and about 120 miles east of Calcutta.

Q.—Is there any thing remarkable in the history of this zillah?

A.—Yes, in the year 1584 it was laid waste by an awful inundation; which was succeeded by invasions of the Mughs aided by the Portuguese of Chittagong: and in the year 1822, another frightful flood occurred, sweeping away houses; and every thing else, and drowning upwards of 37,000 of the inhabitants, and 89,000 head of cattle. From the effects of these repeated catastrophes, it has scarcely yet recovered.

TIPPERA.

Question.—In what province is this zillah?

Answer.—In the province of Bengal; and of which it forms the chief eastern boundary.

Q.—Between what degrees of latitude does it lie?

A.—Between the 22nd and 24th degrees of north latitude.

Q.—What are its boundaries?

A.—Sylhet and Dacca-Jelalpoor on the north; Chittagong and the sea on the south; the Tippera hills and the Burman empire on the east; and the great river Megna, and the district of Dacca-Jelalpoor on the west.

Q.—Has this zillah any other name?

A.—It is sometimes called Roushenabad.

Q.—What may be the extent of this zillah?

A.—In the year 1784, it was estimated at 6,618 square miles, but various lands have since been added; and owing to the wild uncultivated state of the country on its eastern side, its boundary on that side has not yet been positively fixed.

Q.—What number of inhabitants does it contain?

A.—According to the most correct statements that have as yet been made, about 750,000, in the proportion of ten Hindoos to six Mahommedans.

Q.—Are there any rivers of note in this zillah?

A.—The Goomty, which springs from an unknown source, in the eastern mountains, is the only one deserving notice; and it becomes very shallow about 50 miles beyond the town of Comillah.

Q.—What is the name of the capital of this zillah?

A.—Comillah, a town situated in lat. 23° 28′ N., long. 90° 43′ E.; and 51 miles S. E. from Dacca.

Q.—Was this the ancient capital of the Tippera country?

A.—No; about six miles west from Comillah are the remains of many large brick buildings, and of a fort about 200 feet square, the vestiges of the residence of the ancient Tippera Rajahs.

Q.—Are there any other towns in this zillah?

A.—Yes; Noomugur, Luckypoor, Chandpoor, and some others of minor note.

Q.—What are the productions of this zillah?

A.—Cotton, rice, and betel-nut form its most valuable agricultural produce; besides which, it exports large quantities of a coarse but durable description of cotton goods, known all over the world by the names of *baftaes, cossaes.* There are also numbers of wild elephants in this zillah.

CHITTAGONG.

Question.—In what province is the zillah of Chittagong?

Answer.—In the province of Bengal, of which it occupies the south-eastern extremity.

Q.—Between what degrees of latitude does it lie?

A.—Between the 21st and 23d degrees of north latitude.

Q.—What may be its dimensions and soil?

A.—It contains about 2,987 square miles; two-thirds of which comprise an unproductive hilly country, and one-third plain arable land.

Q.—What are the boundaries of this zillah?

A.—The zillah of Tippera on the north; Arracan on the south; the Burmese empire on the east; and the sea on the west.

Q.—What number of inhabitants are there in this zillah?

A.—About 1,500,000; in the proportion of five Mughs and five Hindoos to six Mahommedans.

Q.—What is the name of the capital of this zillah?

A.—Islamabad, also called Chittagong, situated on the west bank of the Chittagong river, about 8 miles from its junction with the sea, and in latitude 22° 22′ N., longitude 91° 42′ E.; about 317 travelling miles from Calcutta. This town is the residence of the Judge and Collector.

Q.—What other towns of note are there in this zillah?

A.—Ramoo, and Cox's Bazar; at both of which places there is a Custom-house. They are chiefly inhabited by Mughs.

Q.—Are there any mountains in this zillah?

A.—The country towards the east and north is rather hilly, and inclined, in some places, to be mountainous; but the blue mountain, called by the Mughs Meindaun, and by the Bengalees Munipahar, which is between 5000 and 6000 feet above the level of the sea, is the only one of any considerable height.

Q.—What are the chief articles of commerce in this zillah?

A.—The exports consist chiefly of timber, planks, canvas, coarse cloth, stockings, and umbrellas; on the sea coast, the Government has a large establishment for the manufacture of salt. And elephants of the best description are caught here in abundance.

Q.—Is there any thing else deserving of notice here?

A.—Yes, a hot well, named Barbacoond, situated at a place called Seetacoond, about 20 miles north of Islamabad.

DISTRICT OF MOORSHEDABAD.

Question.—In what province is this district?

Answer.—In the province of Bengal.

Q.—How many zillahs are there in this district?

A.—Seven; Moorshedabad, Rajeshahee, Beerbhoom, Purneah, Bhagulpoor, Dinajepoor, and Rungpoor.

Q.—In which of these is the circuit court held?

A.—In Moorshedabad.

ZILLAH MOORSHEDABAD.

Question.—What are the boundaries of this zillah?

Answer.—Bhagulpoor and Beerbhoom on the north-west and west; the Pudmawatee river, which separates it from Rajeshahee, on the east; and Nuddea and Jessore on the south.

Q.—What number of inhabitants are there in it?

A.—About 850,000, not including the town.

Q.—What are the names of the principal towns?

A.—Moorshedabad, which is the capital; Bhogobangola, Jungeepoor, Berhampoor, with some others of less note.

Q.—What is the latitude, longitude, and situation of the town of Moorshedabad?

A.—Its latitude is 24° 11′ N., longitude 85° 15′ E., and it is situated on the Bhagirathi, the most sacred branch of the Ganges, about 120 miles north of Calcutta.

Q.—What was its ancient name?

A.—Mucksoorabad.

Q.—Was it ever the capital of Bengal?

A.—Yes, in the year of our Lord 1704, the Nawab Jaffier Khan, (named also Moorsheed Cooly Khan) removed the seat of Government from Dacca here, at which time he changed its ancient name to what which it now has.

Q.—How long did it continue the capital of Bengal?

A.—Until A. D. 1757, when the British made Calcutta the capital.

Q.—Is Moorshedabad a place of much traffic ?

A.—Yes, of very considerable inland trade.

Q.—What may be the population of the town ?

A.—In the year 1814, it was supposed to be about 165,000 souls.

Q.—Is Moorshedabad the residence of any of the public authorities ?

A.—Yes, all the public authorities of the district reside here ; in addition to which there is a court of circuit here.

Q.—Where is Berhampoor ?

A.—About six miles south of Moorshedabad, to which it forms the military cantonment.

Q.—What are the productions of this zillah ?

A.—Silk, indigo, and grain.

Q.—What rivers are there in it ?

A.—The Jallingy intersects it towards the south-east, and the Bhagirathi passes in a southerly direction through its centre : these two streams, being both branches of the Pudma, or Great Ganges, unite at Nuddea, where they assume the name of the Hooghly.

Q.—Are there any canals in this zillah ?

A.—Yes, about thirty-three years ago, a canal was completed between the Ganges and Moorshedabad.

RAJESHAHEE.

Question.—In what province and district is this zillah ?

Answer.—In the province of Bengal, and district of Moorshedabad.

Q.—What are its boundaries ?

A.—Dinajepoor and Mymunsing on the north ; Beerbhoom and Kishnagur on the south ; Dacca-Jelalpoor and Mymunsing on the east ; and Bhagulpoor and Beerbhoom on the west.

Q.—Between what degrees of latitude does it lie ?

A.—It lies between the 24th and 25th degrees of north latitude ?

Q.—In what part of Bengal is it situated ?

A.—Nearly in the centre.

Q.—What number of inhabitants are there in it ?

A.—About 1,500,000, in the proportion of ten Hindoos to six Mahommedans.

Q.—What are the names of the principal towns in this zillah ?

A.—Nattore, Bauleah, Hurriaul, Seebgunge, Coomarcolly, Surda, and some others of less note.

Q.—What are its chief articles of commerce ?

A.—Silk, for which the East India Company have factories at Coomarcolly, Hurnia, Surda, and Bauleah. Cotton is also an article of considerable trade, and is purchased by the Company.

Q.—Are there any rivers ?

A.—Yes, it is much intersected with rivers, and the northern portion, particularly, abounds with large jheels or swamps.

Q.—What is the name of the capital of this zillah ?

A.—Nattore, where the Judge and Collector reside, and the courts are held.

Q.—Where is this town situated ?

A.—In latitude 24° 25′ N., longitude 88° 55′ E., and about 43 miles N. E. from Moorshedabad.

BEERBHOOM.

Question.—In what province and district is this zillah ?

Answer.—In the province of Bengal, and district of Moorshedabad.

Q.—What are its boundaries ?

A.—Bhagulpoor on the north ; Moorshedabad and Nobodeep, or Nuddea, on the east ; the river Aujez, which separates it from Burdwan and Jungul Muhals, on the south ; and Ramgur on the west.

Q.—What is its situation ?

A.—It is situated on the north-western extremity of Bengal, about the 24th degree of north latitude.

Q.—What number of inhabitants does it contain ?

A.—In the year 1801, it was stated to be 1,500,000, since when no estimate has been made ; the proportion then was two Hindoos to one Mahommedan.

Q.—What is the name of the capital town ?

A.—Soory, a town situated in latitude 23° 54′ N., longitude 87° 32′ E., and 50 miles south-west of Moorshedabad. The Judge and Collector reside here.

Q.—What is the nature of the soil in this zillah ?

A.—Hard, and strongly impregnated with iron ; there is also much coal, of so good a quality that the Calcutta

Market is chiefly supplied from it, and much is carried to sea, as ship's stock.

Q.—Are there any iron mines here ?

A.—No ; the soil is rich in iron, but instead of being in veins, as usual, it is mixed in strata, with the clay from which it is extracted.

Q.—Is much of it used ?

A.—Yes, in the neighbourhood of Beerbhoom, Bhagulpoor, and Moorshedabad, particularly ; but it is not so cheap as European iron, which prevents its more general use.

Q.—Are there any forests in this zillah ?

A.—Yes, most extensive ones ; and the power of reproduction rapid.

Q.—What towns may there be in this zillah ?

A.—Besides Soory, there are Nagore, Soorool, Baidyanath, and some others of less note ; Nagore was the capital of the zillah during the Mahommedan sway.

Q.—Are there any rivers ?

A.—None that are navigable, except during the rains: but this is of little consequence to trade, as the materials for road-making are most excellent ; and the roads and bridges are, consequently, better than in most parts of Bengal.

Q.—What are the productions of Beerbhoom ?

A.—The most important are rice, sugar, and silk.

BHAGULPOOR.

Question.—In what province is this zillah ?

Answer.—It occupies a portion of two provinces, Bengal and Behar.

Q.—In what district is it ?

A.—The district of Moorshedabad.

Q.—Between what degrees of latitude does it lie ?

A.—Between the 24th and 26th degrees of north latitude.

Q.—What are its boundaries ?

A.—Tirhoot and Purneah on the north : Ramghur and Beerbhoom on the south ; Purneah and Moorshedabad on the east ; Behar and Ramghur on the west.

Q.—What may be its shape and extent ?

A.—It contains about 8,270 square miles, within an irregular oval, 133 miles long, in a N. N. W. and S. S. E. direction, and 80 miles across its greatest breadth.

Q.—What may be the amount of the population?

A.—Upwards of 2,000,000, in the proportion of three Hindoos to one Mahommedan.

Q.—What is the name of the capital?

A.—Bhagulpoor is the name of the modern capital; and it is situated in latitude 25° 13′ N., longitude 80° 58′ E., and about 110 miles N. W. of Moorshedabad.

Q.—Is Bhagulpoor a large town?

A.—Yes, a large but straggling, ill-built place, containing about 30,000 inhabitants, chiefly Mahommedans. The Judge, Collector, and other public authorities reside here.

Q.—Is there any thing worthy of notice here?

A.—Yes, here is a Roman Catholic Church, and a Mahommedan College, but now in a state of great decay; and about a mile N. W. from the town two remarkable round towers, supposed to be of Jain origin and held very sacred.

Q.—What may be the nature of the soil and appearances of the country in this zillah?

A.—It may be divided into three districts; 1st, that which is subject to the periodical floods of the Ganges; 2nd, that which is sufficiently elevated to be exempt from them, but only gently rising; and 3rd, mountainous.

In the latter, many parts are so rocky, as to be totally incapable of cultivation; in the other two the soil varies much, but on the whole the portions of it fit for the plough, are rich, and capable of being rendered productive.

Q.—What are the productions of this zillah?

A.—Indigo, cotton, rice, barley, wheat, potatoes, pulse, tobacco, betel, and salt; to which may be added, firewood, timber, coal, wax, and honey, which are brought in large quantities from the hills, by the mountaineers, and exchanged for various articles; and small quantities of silk, cloth, and saltpetre which are purchased by the East India Company.

Q.—What towns of consequence are there in this zillah?

A.—Monghyr, formerly a place of some consequence, but now merely an invalid station; Rajmahal, containing about 25,000 inhabitants; Champanagur, and several others of less note.

Q.—What rivers are there in this zillah?

A.—The Ganges is the principal one, but there are

many others which intersect it; some of which are navigable only during the rains, and many, although of considerable width, are fordable at all times, and dry during a great part of the year.

PURNEAH.

Question.—In what province and district is this zillah?

Answer.—Partly in the province of Bengal, and partly in Behar; but in the district of Moorshedabad.

Q.—What are its boundaries?

A.—The Morung hills on the north; Bhagulpoor and the Pudmawatee river on the south; Dinagepoor on the east; and Tirhoot on the west.

Q.—What number of inhabitants are there?

A.—In the year 1810 there were stated, after a careful survey, to be 2,904,380 souls, in the proportion of nine Hindoos to seven Mahommedans.

Q.—What are the names of the principal towns?

A.—Purneah, Nautpoor, Krishbah, Dambdaha, and Matauli.

Q.—Which of these is the capital?

A.—Purneah, a town situated in latitude 25° 45′ N., longitude 88° 23′ E.; and 125 miles N. W. by N. from Moorshedabad. This town occupies a space of nine square miles, and contains only 40,000 inhabitants. The civil authorities of the East India Company reside here, and the courts are likewise held at it.

Q.—What rivers may there be in this zillah?

A.—The largest are the Mahananda, the Coosee, and the Conka, which rise in the Nepaul hills, and flowing through this zillah, pursue their course to the eastward, until they join the Ganges.

Q.—Are the productions of this zillah valuable?

A.—The advantages of soil and climate are so great, that this has always been considered the most productive zillah in Bengal.

Q.—What is the prevailing soil?

A.—Those lands that are overflowed by the Mahananda and its branches, are, upon the whole, rich and fertile; whilst those inundated by the Coosee, especially towards the north and east, are sandy and poor.

Q.—What is the soil in those parts not subject to inundation?

A.—Partly clay, partly free soil, and partly sandy. In a favourable season, the higher lands are productive in grain of all kinds, and a variety of plants, resembling the mustard plant, raised for oil.

Q.—What are the productions of this zillah?

A.—Rice, and every other sort of grain common to Bengal; various plants raised for oil, which is the staple commodity of the zillah; indigo, other dyeing plants, and potatoes.

Q.—Are cattle abundant here?

A.—Yes, the herds of cattle and buffaloes are so numerous, that a considerable quantity of ghee, or clarified butter is annually exported.

Q.—Are there any forests here?

A.—Yes, the northern part, bordering on the Morung, is covered with immense forests of saul and other timber, which being felled, are, during the rains, floated down the rivers to the timber yards in Calcutta.

Q.—What animals and birds are there here?

A.—Wild elephants in great numbers, which are very destructive to the crops; large flocks of pea-fowls, cranes, paroquets, ortalons, and water-hens, which are equally so; immense swarms of water-fowl during the cold season; as also snipes, golden plover, and the lesser bustard, as well as the small white heron, stags, and water crows; in some parts wild hogs, and deer, and a few tigers.

DINAJEPOOR.

Question.—In what province is this zillah?

Answer.—The province of Bengal.

Q.—In what district is it?

A.—Moorshedabad.

Q.—What are the boundaries of this zillah?

A.—Rungpoor and Purneah on the north; Rajeshahee and Mymunsing on the south; Rungpoor and Mymunsing on the east: and Purneah and Bhagulpoor on the west.

Q.—Between what degrees of latitude is it situated?

A.—Between the 25th and 26th degrees of north latitude.

Q.—What are the names of the principal rivers?

A.—The Mahananda, Jamuna, Korotoya, and Tishta.

Q.—What towns are there in this zillah?

A.—Dinajepoor, Malda, Rajgunge, and Bhobaneepoor, are the principal ones.

Q.—Which of these is the capital?

A.—Dinajepoor; containing about 30,000 inhabitants, and situated in lat. 25° 37′ N., long. 88° 43′ E., and about 102 miles N. N. E. from Moorshedabad.

Q.—What number of inhabitants are there in this zillah?

A.—According to the last estimate that was made, and which was in the year 1808, about 300,000, in the proportion of eleven Mahommedans to five Hindoos.

Q.—What are the productions of this zillah?

A.—Rice is the staple commodity; some indigo and sugar are also cultivated.

RUNGPOOR.

Question.—In what province is the zillah of Rungpoor?

Answer.—The province of Bengal, of which it forms the north-eastern extremity.

Q.—In what district is it?

A.—The district of Moorshedabad.

Q.—What are its boundaries?

A.—Bhootan on the north; Mymunsing and the Garrow mountains on the south; Assam and the Garrow mountains on the east; and Dinajepoor on the west.

Q.—Between what degrees of latitude does it lie?

A.—Between the 25th and 26th degrees of north latitude.

Q.—What number of inhabitants does it contain?

A.—About 2,700,000, half Hindoos and half Mahommedans.

Q.—What towns are there in this zillah?

A.—Rungpoor, which is the capital, situated in lat. 25° 43′ N., long. 89° 22′ E., containing about 20,000 inhabitants, and the residence of the Judge and Collector: besides this, there are Mungulhaut and Goalpara.

Q.—What rivers are there?

A.—The Bramhapootra, the Tishta, (or Trisrota,) the Corotoya, and some smaller ones.

Q.—Are there any forests here?

A.—Yes, towards the east of this zillah there are stately forests, but much encumbered with enormous climbing plants, and an undergrowth of weeds, which causes much labour and difficulty in felling the trees.

Q.—What is the nature of the soil, and general appearance of the country?

A.—The soil is generally inferior to that of Dinajepoor: to the eastward the country divides itself into numerous detached ranges of hills, about 8 miles long and 2 broad each,—and, with the exception of the Garrows, not exceeding 1,200 feet in height; the land between these ranges is low, and subject to inundation during the rains.

Q.—What are the productions of this zillah?

A.—Tobacco, and betel-leaf may be considered as the staple commodities of the zillah: wheat is also cultivated in considerable quantities, as well as of late years ginger; quantities of the latter of which are sent to the Calcutta market: barley, cotton, and sugar, are also grown but in small quantities.

Q.—What other articles of produce are there?

A.—Indigo and the poppy; and the farmers rear the mulberry, and the resinous silk-worm, and the lac insect: here are great numbers of wild elephants, and the rhinoceros is not uncommon in the forests; there are also huge tigers and black bears in considerable numbers.

PROVINCE OF BEHAR.

Question.—In what part of India is this province?

Answer.—Behar is a large province in north Hindoostan, lying between the 22nd and 27th degrees of north latitude.

Q.—What is the extent of this province?

A.—It is computed at 26,000 square miles.

Q.—What are its boundaries?

A.—Nipal on the north, the province of Gundwana on the south; Bengal on the east; and Allahabad, Oude, and Bundlecund on the west.

Q.—How is Behar divided?

A.—Within the boundaries above stated, it is divided, naturally, into two equal portions, north and south, by the river Ganges, which here flows in an easterly direction for 200 miles.

Q.—What is the soil and the nature of the country in these two divisions?

A.—The appearance of the country is flat; the soil is the most fertile, and highly cultivated in Hindoostan.

Q.—Is there any other district belonging to this province?

A.—Yes, there is to the southward of these a hilly district, comprehending 18,000 square miles, and including the modern subdivisions of Palamour, Ramghur, and

Chuta Nagpoor; but the value of this district is comparatively small.

Q.—What are the productions of this province?

A.—Opium is its peculiar produce, and the staple commodity of the country. Saltpetre is also manufactured in the division of Hajypoor, and Saum. Cotton is likewise much cultivated, and large quantities of cotton cloth manufactured for exportation.

Q.—What other productions are common?

A.—Grain, sugar, indigo, oil, betel-leaf, rose-water, and otter of roses.

Q.—What is the climate of Behar?

A.—Excepting during the prevalence of the hot winds, comparatively speaking, temperate; and in the winter, cold and bracing; the thermometer ranging in the morning from 35° to 40° Fahrenheit.

Q.—What are the principal rivers of this province?

A.—The Ganges, the Sone, the Gunduck, the Dummodah, Caramnassa, and the Dewah; besides which there are innumerable smaller streams.

Q.—What are the names of the chief towns?

A.—The chief ones are Patna, Chuprah, Daoudnagur, Gaya, Bhagulpoor, Monghir, Arrah, Chittra, and Muzufferpoor. Behar was, at a very remote period, the capital of this province, but has now fallen into great decay.

Q.—Which of these is the capital?

A.—Patna, situated in lat. 25° 27′ N., long. 85° 15′ E., on the south side of the river Ganges, and 400 miles from Calcutta.

Q.—What is the population of Patna?

A.—In A. D. 1811, the city itself was estimated to contain about 312,000 souls, besides sepahees, camp-followers, &c., and the city, with the district attached to it, comprising a circuit of 20 miles, and 909,270 souls, in the proportion of rather more than 2 Hindoos to 1 Mahommedan.

Q.—What districts belong to the Patna circuit?

A.—The Patna division of the court of appeal and circuit comprehends the districts of Ramghur, Behar, Tirhoot, Sarun, Shahabad, and the city of Patna.

Q.—Are there any seminaries in which Hindoo or Mahommedan law is taught here?

A.—No, the students in these branches are instructed by private tutors.

Q.—Are there any religious buildings here?

A.—The Roman Catholics have long had a Church here, and of late years one belonging to the establishment has been built. The ancient Hindoo and Mahommedan religious buildings are falling into decay, without modern ones rising to supply their places.

Q.—Are the natives of the province of Behar, the same as in Bengal?

A.—In stature and appearance they are superior, but much more addicted to the use of intoxicating drugs; and by no means so cleanly in their houses, their persons, or their habits.

Q.—Are they rigid in their religious observances?

A.—Not nearly so much so as in Bengal, and the priesthood of their chief place of worship, Gaya, is the most depraved and degraded in Hindoostan.

Q.—Where is the city of Gaya situated?

A.—In the province and district of Behar, of which latter it is the modern capital, in lat. 24°49′ N., long. 85° E. and 55 miles south from Patna.

Q.—What is Gaya celebrated for?

A.—For being the birth-place of Buddha, the great prophet and legislator of Eastern Asia.

Q.—Are there many Buddhists in this province?

A.—No, the race is extinct, but Gaya is still a revered place of pilgrimage.

Q.—Is there any thing worthy of note in the neighbourhood of Gaya?

A.—Yes, in the adjacent hills are many vaulted caverns cut out of immense masses of solid granite, the interior surface of which has been highly polished.

Q.—Are there any images, or other mythological decorations in these?

A.—No; but on the ground, not far from the base of one of them, called Ramsilla hill, is an insulated figure of Buddha, in a sitting position, and of superhuman size.

PROVINCE OF ALLAHABAD.

Question.—What is the situation of this province?

Answer.—It is a province of Hindoostan Proper, and is situated between the 24th and 26th degrees of north latitude.

Q.—What are its boundaries?

A.—Oude and Agra on the north; the Hindoo province of Gundwana on the south; the provinces of Behar and

Gundwana on the east; and Malwa and Agra on the west.

Q.—What may be its estimated length and breadth?

A.—270 miles its length, and 120 miles its average breadth.

Q.—Into how many districts is it divided, and what are their names?

A.—Eight, viz. Allahabad, Benares, Mirzapoor, Juanpoor, the Rewah territory, Bundlecund, Cawnpoor, and the Manikpoor territory.

Q.—What rivers are there in this province?

A.—The Ganges, Jumna, Goomty, and Caramnassa, with their branches; besides innumerable smaller ones.

Q.—What are the names of the chief towns?

A.—Allahabad, Benares, Callinjer, Chatterpoor, Juanpoor, Mirzapoor, Chunar and Ghazeepoor.

Q.—What may be the appearance of the country, and nature of the soil, in this province?

A.—Near the Ganges and Jumna the country is flat, and the soil very productive; but to the south-west in the Bundlecund district, the country forms an elevated table-land, diversified with high hills.

Q.—What mountains are there in this province?

A.—There are three parallel ranges of hills in the Bundlecund district, each buttressing a table-land, one above the other, and designated by the name of Ghauts.

Q.—What are the names of these ranges?

A.—The first is called the Bindhyachal, or Vindhya mountains. The second called the Pannah Ghauts, seen parallel to these, preserving a distance of about 10 miles. The third, named the Baudair, is the most elevated part of the province.

Q.—What height are these mountains?

A.—The highest summit of the Bindhyachal range does not exceed 2,000 feet above the level of the sea; the heights of the other two have not been ascertained.

Q.—Are there any mines of quarries in these mountains?

A.—Yes, the diamond mines of Pannah in Bundlecund, and the stone quarries of Chunar and Mirzapoor.

Q.—What are the productions of this province?

A.—Wheat, rice, oats, barley, and various descriptions of pulse; also sugar, indigo, and other dyeing plants; flax, cotton, ginger, &c.

Q.—What are the principal exports?

A.—Sugar, indigo, cotton, cotton cloths, opium, diamonds, saltpetre, &c. &c.

Q.—Which is the most celebrated city in this province?

A.—Benares, (called in Sanscrit Varanashi,) both on account of its size, and its being the chief seat of Brahminical learning.

Q.—What is the geographical position of this city?

A.—It is situated in lat. 25° 30′ N., long. 83° 1′ E., and 460 miles from Calcutta.

Q.—What is the population of Benares?

A.—About 200,000 souls; in the proportion of 1 Mahommedan to 3 Hindoos.

PROVINCE OF OUDE.

Question.—What is the geographical position and size of the province of Oude?

Answer.—Oude is one of the smallest provinces of Hindoostan Proper, being about 250 miles long, by an average breadth of 100, and is situated between the 26th and 28th degrees of north latitude.

Q.—What are its boundaries?

A.—On the north, various petty states, tributary to Nepaul, from which it is separated by a range of hills and forests: on the south, Allahabad; on the east, Behar; and on the west, Delhi and Agra.

Q.—What may be the appearance of the country, and the nature of the soil?

A.—The whole surface of this province is level, well watered with rivers, and, when well cultivated, extremely productive.

Q.—What rivers are there in Oude?

A.—The Ganges, (which bounds it to the west,) the Goggra, and the Goomty, are the principal ones; but there are many smaller streams.

Q.—What are the subdivisions of this province?

A.—Lucknow, Fyzabad, Khyrabad, Baraitche, with a section of Manikpoor, (composing the reserved dominions of the king of Oude,) and Goruckpoor, belonging to the British.

Q.—What is the name of the capital city of Oude?

A.—Lucknow, situated on the south side of the Goomty in lat. 29° 50′ N., long. 80° 50′ E., and 650 miles from Calcutta.

Q.—What other towns are there?

A.—Fyzabad, Oude, Khyrabad, Baraitche, and Tanda are the principal ones.

Q.—What are the productions of this province?

A.—Wheat, barley, rice, and other grains; sugar-cane, indigo, poppies for opium, and all the richest articles raised in India; to which we may add, as a mineral production the Lapis Lazuli.

PROVINCE OF AGRA.

Question.—What is the geographical situation and size of the province of Agra?

Answer.—Agra is a large province of Hindoostan Proper, situated between the 25th and 28th degrees of north latitude, about 250 miles in length, by an average breadth of 180.

Q.—What are its boundaries?

A.—Delhi on the north; Malwa on the south; Oude and Allahabad on the east; and Ajmeer on the west.

Q.—What may be the general appearance of the country and nature of the soil?

A.—To the north-east of the Jumna, the surface is flat and open; but to the south of the Chumbul, and towards the western frontier, it is more hilly and jungly; the soil is, generally speaking, of a light sandy description.

Q.—What rivers are there in this province?

A.—The Ganges, Jumna, and Chumbul; and many smaller ones, such as the Sindee, Kohaez, &c.

Q.—What are the number and names of the subdivisions of Agra?

A.—Eight, viz. 1, Agra; 2, the Doab; 3, Etawah; 4, Furruckabad; 5, Calpee, Gohud, and Gwalior; 6, the Bhurtpoor Territories; 7, Alvar and Machery; and 8, Allighur.

Q.—What is the name of the capital of the province?

A.—Agra; situated on the south-west side of the Jumna, in lat. 27° 11′ N., long. 77° 54′ E., and 950 miles from Calcutta.

Q.—What other towns are there?

A.—Alvar, Bhurtpoor, Deeg, Mathura, Bindrabund, Kanoje, Etawah, Gwalior, Gohud, Calpee, Narwar, and Furruckabad.

Q.—Of what description are the inhabitants?

A.—In general a robust, handsome race of people, consisting of Hindoos and Mahommedans.

Q.—What languages are used in this province?

A.—The Hindoostanee is the vulgar tongue; but the Persian is spoken by the higher class of Mahommedans, and is used in all official documents. The Bruj dialect is spoken immediately round the city of Agra, and extends to the Vindhya mountains.

Q.—What are the productions of Agra?

A.—Indigo, cotton, and sugar; and the horses are considered superior to any raised in the east, excepting such as are reared in the Government studs.

Q.—What may be the population of the province?

A.—According to the most correct statements that have as yet been obtained, it does not exceed 8,000,000.

Q.—What is the prevailing religion?

A.—The Hindoo.

Q.—Is there any thing worthy of note regarding the city of Agra?

A.—Yes, it is supposed to have been the birth-place of the Avatar, or incarnation of Vishnu, under the name of Parasu Rama, whose conquests extended to, and included, Ceylon.

PROVINCE OF DELHI.

Question.—Where is this province situated?

Answer.—Delhi is an imperial province of Hindoostan Proper, situated between the 28th and 31st degrees of north latitude.

Q.—What are its boundaries?

A.—Lahore and North Hindoostan, west of the Goggra, on the north; Agra and Ajmeer on the south; Oude and North Hindoostan on the east; and Ajmeer and Lahore on the west.

Q.—Into how many districts is Delhi divided, and what are their names?

A.—Ten, viz. the assigned territories, Bareilly, Mooradabad, Shajehanpoor, Rampoor, North Saharunpoor, South Saharunpoor, (or Meerut,) Hurriana, Sirhind, and Pattialah, and various petty Seik states.

Q.—What rivers are there?

A.—The Ganges, Jumna, Caggur, Chittung, and the almost extinct Sereswati.

Q.—What is the nature of the soil?

A.—An arid sand, but capable of being brought into a very productive state, if by means of canals a sufficient quantity of water were reserved for the purposes of irrigation.

Q.—Are there any canals in this province?

A.—Yes, the canal of Ali Merdan Khan, extending from the river Jumna opposite to Kurnaul to Delhi, through a winding course of 180 miles, besides some lateral branches: it had been long choked up, but was re-opened by order of Government, in A. D. 1820, and is the only source of supply for pure drinking water to the inhabitants of that city.

Q.—Are there any others?

A.—Yes, the canal of the Sultan Firoze Shah, which separates from that of Delhi a little below Kurnaul, and stretches to the westward through the Hurriana, by Hansi and Hissar, to the frontiers of Biccaneer; and Zabeta Khan's, or the great Doab canal, which leaves the Jumna a few miles below its issue from the northern mountains, and after a course of 150 miles, joins it again opposite Delhi.

Q.—Are these two canals sufficiently open to be useful?

A.—No; they are completely choked up.

Q.—Are the inhabitants well supplied with water?

A.—No; in the dry season they are obliged, in most places, to dig from 120 to 200 feet, and then only obtain brackish water.

Q.—Is the province of Delhi well cultivated?

A.—No, owing to the want of canals, which might easily be dug, the surface of this province, generally speaking, is in a state of naked sterility, unless immediately after the rains, when, for a short time, there is good pasture.

Q.—What are the productions of this province?

A.—Between the Jumna and the Sutuleje mango trees are numerous; and wheat, barley, gram and other grains are also cultivated: but even in this, which is the least arid of all the districts, irrigation is necessary when the water is not within ten or twelve cubits of the surface.

Q.—What are the chief towns?

A.—Delhi, Bareilly, Pillibeet, Shahjehanpoor, Rampoor, Mooradabad, Anopshehr, Meerut, Sheerdhana, Saharunpoor, Pattialah, Ambalah, and Sirhind.

Q.—Which of these is the capital?

A.—Delhi, the ancient capital of the Patan and Mogul empires, situated in lat. 28° 41′ N., long. 77° 5′ E.

Q.—What is the size of this city?

A.—Its ruins occupy a space of nearly twenty square miles: the present town occupies a space of about seven miles in circumference.

Q.—Are there any objects worthy of notice in this ancient city?

A.—Altogether, the ruins present the most striking scenes of desolation to be met with throughout the whole world. Some of the gates, caravanserais, and mosques are still tolerably entire.

Q.—What distance is Delhi from Calcutta?

A.—By the Beerbhoom road 976 miles.

PROVINCE OF LAHORE.

Question.—In what part of Hindoostan is Lahore situated?

Answer.—Lahore is a large province of Hindoostan Proper, extending from the 30th to the 34th degrees of north latitude, about 340 miles in length, by an average breadth of 200 miles.

Q.—What are its boundaries?

A.—The province of Cashmeer, and the course of the Indus, on the north; Delhi, Ajmeer, and Mooltan, on the south; the mountains of northern Hindoostan on the east; and the Indus, which separates it from Afghanistan on the west.

Q.—What are the sub-divisions of this province?

A.—It naturally forms two grand divisions, the one denominated the Punjab, or flat country; the other, the Kohistan, or hill country.

Q.—What are the sub-divisions of the Punjab?

A.—There are five, viz. the Doabeh Sinde Sagor, the Doabeh Jinhut, the Doabeh Rechna, the Doabeh Bauz, and the Doabeh Jallinder.

Q.—What are the sub-divisions of the Kohistan?

A.—There are ten, viz. Kishtewer, Chandahnee, Jamboe, Kangrah, Cooloo, Chamba, Mundi, Sukait, Hurpoor, and the Gurkers.

Q.—To whom is this province subject?

A.—The whole of the first, and a large portion of the last, is either directly subject, or tributary to the Rajah of Lahore.

Q.—What is the climate of Kohistan?

A.—In summer, warm, but in winter the cold is about the same degree as that experienced in the central regions of Europe.

Q.—What are the productions of Kohistan?

A.—Wheat, barley, and a variety of small grains; and a small quantity of rice.

Q.—What are the productions of the Punjab?

A.—Wheat, barley, rice, pulse of all sorts, sugar-cane, tobacco, and various fruits.

Q.—What are the chief rivers of the province of Lahore?

A.—The Indus, Sutuleje, Beyah, Ravi, Chinaub, and Jhylum.

Q.—What is the name of the capital city of this province?

A.—Lahore, a city situated on the south side of the river Kavery, in lat. 31° 36′ N., long. 74° 3′ E., and 1,356 miles from Calcutta.

PROVINCE OF CASHMERE.

Question.—Where is Cashmere?

Answer.—It is a province of Hindoostan Proper, lying between the 34th and 35th degrees of north latitude.

Q.—What are its boundaries?

A.—It is surrounded by lofty mountains, which separate it from Little Tibet, on the north; from Ladack, on the east; from Lahore, on the south; and Puckely on the west.

Q.—What are its dimensions?

A.—Including the mountains, it is about 110 miles in length, by an extreme breadth of 60; being in shape a broad oval.

Q.—How is Cashmere entered?

A.—By seven passes through the mountains, viz. four from the south, one from the west, and two from the north. The pass of Bember is the best; but that of Muziffarebad is most frequented.

Q.—What are the productions of Cashmere?

A.—Rice, wheat, barley, and various other grains; as well as most of the plants, fruits, and flowers, common in Europe: a superior sort of saffron is cultivated in the plains, and iron, of an excellent quality, is found in the mountains.

Q.—What rivers are there?

A.—The Jhylum, or Hydaspes, is the largest; but there are innumerable mountain streams.

Q.—What are the chief articles of commerce?

A.—The Cashmere shawl, of which it is calculated there are about 80,000 manufactured annually; a peculiar kind of writing paper, said to be the best in the East, was formerly an article of great traffic, as were also lacquered-ware, cutlery, and sugar; but trade of all sorts has, for many years, been very languid.

Q.—Of what description are the inhabitants of Cashmere?

A.—A stout well-formed people, of a dark brunette complexion; lively, acute, keen in the pursuit of wealth, and proverbial liars.

Q.—What number of inhabitants are there?

A.—They have been calculated at between 500,000 and 600,000; but their number is very imperfectly known.

Q.—What is the name of the capital?

A.—Cashmere, (*or Serinagur,*) a dirty ill-built town; situated in lat. 33° 23′ N., 47° 47′ E., and extending about three miles on each side of the Jhylum, over which there are four or five wooden bridges. It was in A. D. 1809 said to contain from 150,000 to 200,000 inhabitants.

PROVINCE OF AJMEER, OR RAJPOOTANA.

Question.—Where is Ajmeer situated?

Answer.—Ajmeer is a large province of Hindoostan Proper, situated between the 24th and 31st degrees of north latitude, in length about 350 miles, by an average breadth of 200.

Q.—What are its boundaries?

A.—Mooltan, Lahore, and Delhi, on the north; Gujerat and Malwa on the south; Delhi and Agra on the east; an Mooltan on the west.

Q.—Has it any other name?

A.—Sometimes it is called Marwar, but this appellation belongs properly only to the Joudpoor territories.

Q.—What are the sub-divisions of Ajmeer?

A.—They are fourteen in number, viz. the Bhaty country, Biccaneer, the great sandy desert, Jesselmere, Joudpoor, Marwar, Nagore, Shekawatty, Jeypoor, Ajmeer district, Harowty, Odeypoor, Mewar, and Sarowz.

Q.—What is the nature of the soil?

A.—Arid sand of so saline a quality as spontaneously to produce much nitre: a very large portion of it is destitute of inhabitants, water and vegetation; the only vegetable to be found being the water-melon, which in the dryest parts is met with in great profusion, and of a very large size; this desert is in some places 400 miles broad extending far beyond the boundaries of Rajpootana.

Q.—What articles of produce and animals are there?

A.—It produces grain sufficient, in ordinary years, for its inhabitants; and camels are numerous, as also bullocks; among the wild animals are the antilope, and the Goorkur or wild ass, mentioned in the book of Job.

Q.—What towns are there?

A.—Ajmeer, Neemuch, Nusseerabad, and some others.

Q.—Which of these is the capital?

A.—Ajmeer, situated in lat. 26° 31′ N., long. 74° 28′ E., and 1,030 miles from Calcutta; was formerly the capital; but in A. D. 1823, Neemuch was formed into a residency, and in 1826, Sir John Malcolm recommended it to be selected as the head-quarters of the public authorities, on account of its equal distances from the capitals of the principal Rajpoot and Malwa chieftains.

PROVINCE OF MOOLTAN.

Question.—Where is Mooltan?

Answer.—Mooltan is situated between the 24th and 30th degrees of north latitude, and was one of the largest provinces of the ancient empire of Hindoostan, but it is now very much reduced in size.

Q.—What are its present boundaries?

A.—Lahore on the north; the Indian Ocean on the south; the river Indus on the west; and the great desert of Ajmeer on the east.

Q.—What are the modern divisions of this province?

A.—Mooltan Proper, Bahawulpoor, Backar, Sinde, Tatta, and Chalchkaun.

Q.—What is the nature of the soil?

A.—Generally speaking, an arid and unproductive sand, destitute of grass, but scattered over with hardy bushes.

Q.—Are there no fertile districts?

A.—Mooltan Proper, and Bahawulpoor, which lie on the banks of the Indus, and within the reach of its periodical inundations, are fertile; and here, as also immediately round the villages, which are placed near deep water-courses, wheat, cotton, and corn, are to be seen.

Q.—Is this province subject to British dominion?

A.—No, it is far removed from the British territories; possesses neither political nor commercial importance, and is seldom visited by Europeans.

Q.—What is the name of the capital?

A.—Mooltan, a city situated in lat. 30° 9′ N., long. 71° 7′ E., surrounded by a wall, from 40 to 50 feet high, with towers at regular distances, and having a citadel on a rising ground, and several fine tombs.

Q.—What are the chief productions of this city?

A.—Mooltan is noted for its silks, and a species of carpet much inferior to those of Persia. The general appearance of the city and environs, is, that of a country once prosperous, but now in a state of decay.

Q.—Is this a very ancient city?

A.—Yes, it is supposed to have been the Malli of Alexander's historians, and is spoken of, in A. D. 1582, as one of the most ancient cities of Hindoostan.

PROVINCE OF CUTCH.

Question.—Where is this province?

Answer.—It is principally situated between the 23rd and 25th degrees of north latitude, and consists of two portions; the Runn, an immense salt morass; and an irregular hilly district, surrounded by the Runn and the sea.

Q.—What are its boundaries?

A.—The sandy desert on the north: the Indian Ocean on the south; the Gulf of Cutch on the east; and the easternmost branch of the Indus, and a barren waste on the west.

Q.—What is the appearance of the country?

A.—It is almost entirely covered with volcanic matter; the rocks appear to have been split by the action of fire, and their vicinity abounds with specimens of metallic scoria: and through the centre of the province runs a range of moderate-sized mountains, named the Lakhi, which are destitute of both soil and water.

Q.—Are there any rivers?

A.—None, although numerous mountain torrents during the rains, which are nearly dry after they cease.

Q.—What is the nature of the soil?

A.—A light clay, covered to the depth of about five inches with a coarse sand. Iron ore is found every where, and a species of wood-coal of a tolerably good quality, about twenty feet below the surface. Bituminous earths, and ligneous petrifactions also abound.

Q.—Does Cutch produce grain sufficient for its own consumption ?

A.—No, a large portion of what is required is imported from Gujerat, Malabar, and Sinde, in return for which cotton is principally exported.

Q.—What are the names of the principal towns ?

A.—Bhooj, Mandavia, Luckputbunder, Moondra, Anjar, Koteser, Nangercha, Kotara, Roha, Abera, Sandhan, Kyra, Mothara, Rowpoor, Adooi, Wandea, and Anysir.

PROVINCE OF GUJERAT.

Question.—Where is this province ?

Answer.—Gujerat is a large province of Hindoostan Proper, situated between the 21st and 24th degrees of north latitude.

Q.—What are its boundaries ?

A.—Ajmeer on the north ; the Indian Ocean and Aurungabad on the south ; Malwa and Candeish on the east ; and Mooltan, Cutch, and the sea, on the west.

Q.—What are the divisions of this province ?

A.—Puttenwara, Jutwar, Gujerat, Peninsula, Chowal, Ederwara, Wagur, Chumpaneer, Cherooter, Ahmedabad, Baroda, Broach, Nandole, Surat, and Attaveesee.

Q.—What is the general appearance of the country ?

A.—There is a steep and craggy range of mountains on the north-east ; but the general aspect of the country is flat, and nearly the whole of the south-west division is an open fertile plain : towards the west boundary the country is either level and arid, or a low salt swamp.

Q.—What rivers are there ?

A.—Gujerat is traversed by several noble rivers, viz. the Nerbudda, Tuptee, Mahy, Mehendey, and Sabirmatty ; but, notwithstanding, there is a great scarcity of water in some districts.

Q.—What are its principal sea-ports ?

A.—Surat, Broach, Cambay, Bownugger, and Jumboseer.

Q.—What are the names of the principal towns ?

A.—Surat, Ahmedabad, Broach, Baroda, Cambay, Gogo, Bownugger, Chumpaneer, and Junaghur.

Q.—What are the productions of Gujerat.

A.—Horses, excellent bullocks, and draught cattle: coarse cloths, saltpetre, hemp, indigo, and opium, as well as other productions common to tropical climates.

THE DECCAN.

Question.—What is the 3rd grand division of Hindoostan?

Answer.—The Deccan.

Q.—What are the boundaries of the Deccan?

A.—The Nerbudda river, and an imaginary line commencing at its source, and extending in the same parallel of latitude as far as the Hooghly, or western branch of the Ganges, on the north; the rivers Krishna and Toombudra on the south; the Bay of Bengal on the east, and the Indian Ocean on the west.

Q.—Into what number of provinces is the Deccan divided?

A.—Into nine.

Q.—What are their names?

A.—Gundwana, Orissa, the Northern Circars, Candeish, Berar, Beeder, Hyderabad, Aurungabad, and Bejapoor.

Q.—Which of these belong to the British?

A.—Orissa, Candeish, the districts of Ahmednugger, and Poona, in the province of Aurungahad; and the district of Dauvar in the province of Bejapoor.

Q.—What rivers are there in the Deccan?

A.—None that are navigable: when swollen with the rains they are too impetuous, and during the dry season, they are too shallow.

Q.—What is the general character of the country?

A.—The general charater of the north-western quarter of the Deccan, now subject to the British Government, is that of a barren country; but in other parts, very mountainous, and covered with thick forests.

Q.—What are the inhabitants?

A.—Chiefly Hindoos.

INDIA, SOUTH OF THE KRISHNA.

Question.—What is the name of the 4th grand division of Hindoostan?

Answer.—India, south of the Krishna, or as it is more frequently called, the Peninsula of India, although its form is that of an equilateral triangle.

Q.—What are its boundaries?

A.—The river Krishna constitutes its base on the north; Cape Comorin its apex on the south; the Coast of Coromandel, bounded by the Bay of Bengal, on the east; and the Coast of Malabar, bounded by the Indian Ocean, on the west.

Q.—Into how many provinces is it divided?

A.—Into nine.

Q.—What are their names?

A.—Canara, Malabar, Cochin, Travancore, the Balaghaut Ceded Districts, Mysore, Coimbatoor, Salem and Banamahal, and the Carnatic.

Q.—What is the geographical feature of this region?

A.—A central table-land, elevated from 3,000 to 5,000 feet above the level of the sea; separated by wild, abrupt, and mountainous declivities, from the low countries to the east and west, which form a belt, of unequal breadth, between it and the sea.

Q.—What are the names by which these divisions are usually known?

A.—The table-land is usually termed Balaghaut, (above the ghauts;) and the lower belt, Pazenghaut, (below the ghauts.)

Q.—What are the inhabitants of this division?

A.—Chiefly Hindoos; the Mahommedans are, comparatively speaking, few in number.